T0118479

SAVING AMERICA

"More Than One Long Day"

Governmental and Corporation Deception = Patent Thievery

SAVING AMERICA
"More Than One Long Day"

Governmental and Corporation Deception = Patent Thievery

Deceptive Corporations using this patent
and deterring terrorism
"Our claim to locating and the dispatching of Al Qaeda"
Our Patent To Protect

By Daniel B. Hock

Authored and dictated by; Daniel B. Hock-c; 2012
Formatted and prepared by, My Loving Wife; Carol
To whom I dedicate this book

iUniverse, Inc.
Bloomington

SAVING AMERICA "More Than One Long Day" Governmental and Corporation Deception = Patent Thievery

iUniverse books may be ordered through booksellers or by contacting:

iUniverse
1663 Liberty Drive
Bloomington, IN 47403
www.iuniverse.com
1-800-Authors (1-800-288-4677)

ISBN: 978-1-4759-5063-2 (sc)
ISBN: 978-1-4759-5066-3 (ebk)

Printed in the United States of America

iUniverse rev. date: 09/17/2012

INTRODUCTION

Many terrorists are being hunted after. Gadhafi and Bin Laden for instance. There had been many others before and now. With the continuing support of our military and Coast Guard, the dispatching of these heinous monitored felon terrorists, have finally been traceable.

I truly believe, a patented monitoring Auto-ID GPS tracking, device, implemented into cell phones, was a detrimental part of these terrorists being located. This patented device was inception, invented, and fully patented by my wife and I.

We were deceived by American companies and our unit was built giving these companies over $900 million in contracts by our government. This continues to grow.

Non-disclosures with American companies meant nothing. Construction of our patented, cellular GPS auto ID locator device constructed by a cellular company without any approval, or arrangement.

We Are Proud

Abuse of small business,

We Are Proud

In this book, I will relay patent information, inception, description of prototype, and timelines of American companies contact. With your understanding of our belief, I just wonder why a presidential e-mail is all of our thanks.

Our belief in Destiny

This is our book of dreams, innovation, despair, and trusting, than deceived. Our ever-growing insight involving governmental and large American corporations had us believing, in deception.

Every mind is a genius. You must remember the government and corporations have teams of geniuses, only to obscure your dreams and innovations. My mistake cost "our dreams", innovation, all insight and destiny.

For 23 years as a professional Dayton Ohio firefighter, my career was, a life rescuer. Imagine the unbearable screams of terror. The screams, I was dedicated to relieve. My belief was that I COULD make a difference nationally, and protect my family as well. My wife and family is my driven force. This force will not be reckoned with.

A feeling of rescue elation is overwhelming. My dedication to protect has propelled my insight, dramatically. How could this come to me? Why? There was to be a reason. I could feel a destiny to protect, Nationally? Is it my desire to protect globally?

My drive to understand will continue until I am satisfied. Soul heartedly! No American corporation, or government, will stand in my way.

As claimed earlier, my beliefs will always be that our patent has an exceptional value of protection for our nations children, our military, and public safety. Terrorism is becoming a worse concern. With proven technology our country is becoming safer.

My wife and I have personally witnessed mistrust by American corporations. The victims being stalked by pedophile, predators, and court-ordered cases must be addressed. With the entities covered under our patent award, this gives us entitlement to help children and our military, utilizing satellite tracking with auto ID to help control US borders and our troops overseas. This is a global effort.

The year of 2000s had been a very insightful year for us. We have had many dreams, and much despair. The dreams of the happiest, is always with my family. The despair was the worst during rescues, and my insight. Some, with heart wrenching feelings of death, and other disparity, our dealings with deceptive governmental corporations, and others, capitalizing on our, inception, patented innovation and invention.

The American Corporation names and agreements will be eliminated from this book. "These forms cannot be displayed for legal reasons". But, this information is the truth and nothing but the truth. By reading this truth, you may see how over $1.5 billion has already been generated since the inception of our patent. These monies grow daily. This $1.5 billion contract was not given to us but in a contract to an American company, by our government. It also seems our government trends to lean toward agreements with disadvantaged corporations. This deprives small businesses to start up. Isn't this what our dreams and innovations are to be? President Obama? Not attorney cover-ups of every word of context to avert infringement.

We are not a disadvantage Corporation, but as with every start up business, protection steps must be taken to protect you. Unfortunately, our American Corporation was determined to eliminate our involvement, with their protection steps.

Innovation and Inception

My dedication and compassion began when I was hired as a professional life rescuer. Throughout my career, I have rescued many, and lost a few. The few, that were lost, struck my heart with devastation. The fear that gave way was to strongly drive my future destiny. This destiny, I feel, was a wonderful belief that I could achieve a great future.

For over 25 years I have prayed for the strength and knowledge to rescue all, and not lose, any. I was given the strength and adrenaline to drive for that goal. I felt that my destiny would be fulfilled. To this day, I am thankful for the knowledge and drive that I was givèn. The prayers have worked. This is what has given to me my wife, Carol and a wonderful family.

My wife and I have had many setbacks. I believe the setbacks, or a test, were to be our strength to be together. Carol and I set our goals to protect. We have been husband-and-wife for over 19 years. Together we have made a bond to ensure the protection of our children. We have three great sons, with two, soon to be, daughters-in-law.

After 23 years, my life rescue career was ended due to injuries leading to retirement. This life career of rescue has NOT ended. After we moved to a larger populated area, we investigated near our home for predators. Within one square mile we were alarmed of the large amount of pedophiles and free roam, tier 3 predators. These are the worst of the worst. We learned that these predators wore monitored, ankle transmitters. I set out to understand these transmitters.

After locating online an expired monitoring system, my wife located a device that worked to alert you by a beeper that would be placed in a bag or luggage, and would beep when you "stepped away" from your luggage. I would consider reversing this technology. I was also interested in cellular auto ID, with a picture of the felon, and offenses.

My idea was to read the signal from the ankle transmitter to receive information and warnings of approaching danger. A body worn, beeper or silent passive device would be the base model. By having cellular, GPS auto ID, auto locator added, a much larger area of safety can be attained.

I have studied the monitoring companies of these predators, and the ankle monitor. Our prototype was taking shape. During this time while living in Florida, we would worry not only for the safety of just our child, but also for all children.

My wife and I contacted a patent attorney and described our intentions to invent and patent the "system to alert", by RFID and GPS. This patent was accepted as patent pending, in October of 2005. We now have, our destiny to protect. From that time, I contacted many companies of interest. An area circuit company was my 1st contact. They were not interested in this monitored predator safety alert system. I contacted many other sources with a negative outcome.

Our patent attorney who helped with the wording of our patent was not sure of the addition of, cellular GPS auto ID. The full patent was approved with all tactical alerts, passive device alerts, and cellular GPS auto ID locator information.

APPLICATION

FOR THE UNITED STATES LETTERS PATENT

———————

SPECIFICATION

TO ALL WHOM IT MAY CONCERN:

BE IT KNOWN THAT I, DANIEL B. HOCK, a citizen of the United States of America, have invented new and useful improvements in a MONITORED FELON WARNING SYSTEM of which the following is a specification:

A MONITORED WARNING SYSTEM

ABSTRACT OF THE DISCLOSURE

A monitored felon warning system has a police computer having a felon database. The system also has a global positioning system, known as GPS, and a sending subassembly having a circuit, with the transmitter being capable of transmitting a 1st signal to a receiver. The receiver gives off an alarm and also causes a display of warning material, such as a photograph to be displayed on the cell phone screen.

BACKGROUND OF THE INVENTION

Field of the Invention

The present invention relates to a monitored felon warning system and more particularly pertains to monitoring felons and persons under court order.

Description of the prior art

The use of monitoring devices of known configurations is known in the prior art. More specifically, monitoring devices of known configurations previously devised and utilized for the purpose of monitoring the location of persons are known to consist basically of familiar, expected, an obvious structural configurations, notwithstanding the myriad of designs encompassed by the crowded prior art which has been developed for the fulfillment of countless objectives and requirements.

The monitored felon warning system according to the present invention substantially departs from the conventional concepts and designs of the prior art, and in doing so provides an apparatus primarily developed for the purpose of monitoring felons and persons under court order.

Therefore, they can be appreciated that there exists a continuing need for a new and improved monitored felon warning system that can be used for monitoring felons and persons under court order. I would understand that a cellular monitored felon, by use of GPS, could fall into the category of monitoring terrorists. In this regard the present invention substantially fulfills this need.

SUMMARY OF THE INVENTION

In view of the foregoing disadvantages inherent in the known types of monitoring devices of known configurations now present in the prior art. The present invention provides an improved monitored felon warning system. As such, the general-purpose of the present invention, which will be described subsequently and generate greater detail, is to provide a new and improved monitored felon warning system and method which has all the advantages of the prior art and none of the disadvantages.

To attain this, the present invention essentially comprises a monitored felon warning system for allowing a user to get information of approaching to, or from, monitored felons. The system comprises several components, in combination.

First provided is a dedicated master computer. The master computer has a felon database. The computer has a memory and a processor. The computer is electronically coupled to a cellular telephone system. The database of the computer is configured to store information concerning a member of the class of persons that include convicted felons, persons under court order, and other monitored felons deemed dangerous. Each member of the class of persons is identified with a unique identifying code.

The database also is enabled to transmit warning information concerning convicted felons and persons under court order via the cellular telephone system. The warning information transmitted is a member of the class of information that includes physical description, photograph and warnings, including "be-on-the-lookout", or BOLO warnings.

Next provided is a global positioning system, also known as GPS.

Next provided is a sending subassembly. The sending subassembly has a circuit. The circuit comprises a processor, a power source, a transmitter and a global positioning system location identifier. The transmitter is capable of transmitting a 1st signal. The 1st signal contains the unique identifying code and the GPS location of the sending subassembly.

Next provided is a person wearing, carrying, or in the case of, a microprocessor integrated cellular device, nearby, as the sending subassembly. The person as a member of class of persons that include convicted felons, monitored felons, and persons under court order.

A person under court order may include persons who are the subjects of restraining orders to keep a certain distance from another person or a place.

Next provided is a receiving subassembly. The receiving subassembly comprises a power source and a circuit. The circuit comprises a memory, a processor, a receiver, a transmitter, a GPS location identifier and an alarm. The alarm is a member of the class of alarms that include audio alarm, tactical alarm and visual alarm. The receiver subassembly is capable of receiving and processing the 1st signal containing the unique identifying code and the GPS location originating from the sending subassembly. The receiver subassembly is configured to produce a 2nd signal. The 2nd signal comprises the unique identifier, the receiver subassembly GPS location and the sending subassembly GPS location.

Next provided is a cell phone, having an LCD screen. The cell phone has a circuit, an associated processor and an associated memory. The cell phone is also electronically coupled to the receiving subassembly. The cell phone is also electronically coupled to the felon database. The cell phone receives the 2nd signal from the receiving subassembly. The cell phone then couples to the database and transmits the contents of the 2nd signal to the database, where it is processed and stored.

Lastly provided is the database, upon receiving the contents of the 2nd signal, then generates and transmits a 3rd signal containing the felons warning information to the cell phone. The warning information is then displayed on the LCD of the cell phone for the user to view.

There has thus been outlined, rather broadly, the more important features of the invention in order that the detailed description thereof that follows may be better understood and in order that the present contribution to the art may be better appreciated. There are, of course, additional features of the invention that will be described hereinafter and which will form the subject matter of the claims attached.

In this respect, before explaining at least one embodiment of the invention in detail, it is to be understood that the invention is not limited in its application to the details of construction and to the arrangements of the components set forth in the following description or illustrated in the drawings. The invention is capable of other embodiments and of being practiced and carried out in various ways. Also, it is to be understood that the phraseology and terminology employed herein are for the purpose of descriptions and should not be regarded as limiting.

As such, those skilled in the art will appreciate that the concept shouldn't, upon which this disclosure is based, may readily be utilized as a basis for the designing of other structures, methods and systems for carrying out the several purposes of the present invention. It is important, therefore, that the claims be regarded as including such equivalent constructions insofar as they do not depart from the spirit and scope of the present invention.

It is therefore an object of the present invention to provide a new and improved monitored felon warning system, which has all of the advantages of the prior art monitoring devices of known configurations and none of the disadvantages.

It is another object of the present invention to provide a new and improved monitoring felon warning system, which may be easily and efficiently manufactured and marketed.

It is another object of the present invention to provide a new and improved monitor felon-warning system which is of durable and reliable constructions.

And even further object of the present invention is to provide a new and improved monitored felon warning system which is susceptible of a low cost of manufacture with regard to both materials and labor, and which accordingly is then susceptible of low prices of sale to the consuming public, thereby making such a Monitored Felon Warning System economically available to the buying public.

Even still another object of the present invention is to provide a monitored felon warning system for monitoring felons and persons under court order utilizing the cellular GPS auto tracking ability of this patent.

Lastly, it is an object of the present invention to provide a new and improved monitoring felon warning system comprising a policed master computer having a felon database. The system also comprises a GPS and a sending subassembly having a circuit with a transmitter being capable of transmitting a 1st signal to a receiver. The receiver gives off an alarm and also causes a display of warning material, such as a photograph to be displayed and/or information of offenses on the cell phone screen.

These together with other objects of the invention, along with the various features of novelty, which characterize the invention, are pointed out with particularity in the claims annexed to and forming a part of this disclosure.

For a better understanding of the invention, its operating advantages and the specific objects attained by its uses, reference should be had to the accompanying drawings and descriptive matter in which there is illustrated preferred embodiments of the invention.

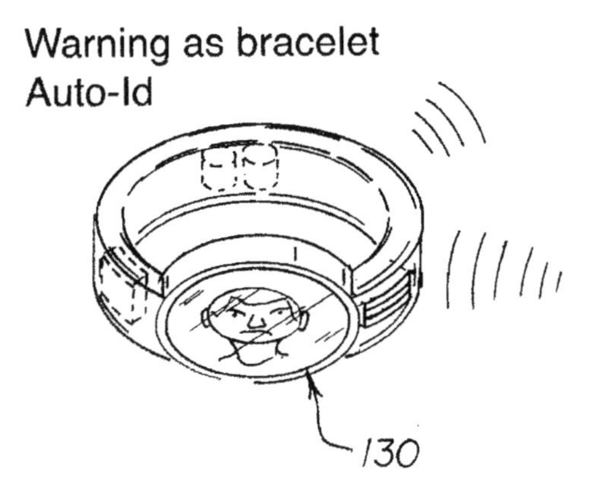

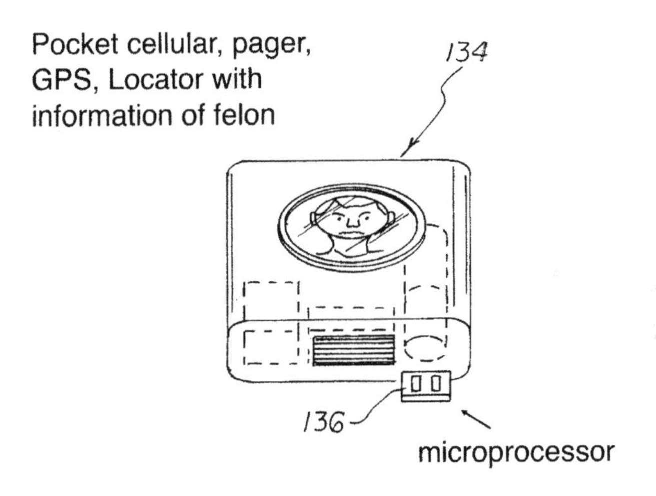

Carol was reading the paper one day and told me of an audition for American Inventor2. My wife decided that I should apply alone. Hundreds of inventors were there. I wish to be one of the lucky ones to pass the addition. I can remember the gentleman with the one-man sand shovel and bagger. There was also another with a 0 turn push lawnmower. When it was my turn to explain our invention, I was quite nervous but did my best. My outcome was not good. The woman that listened and recorded our invention, and intention to protect our children from predators, was not good enough she stated, "of course every parent would be concerned about their child." American inventor2 was not capable of producing, or introducing, our safety product. I do remember that an interest of a car bib, to eat while you drive and not drip food on yourself was more important? The winner of American Inventor2 was the gentleman with the one-man sand shovel bagger. I couldn't believe it. I remember his idea was military bought for 1 million dollars.

That was one of many setbacks, and my learning of DETRIMENT. My drive continued. More business contacts began a long road.

P 207, 6

Inventor or Team Name: DANIEL B. HOCK

"AMERICAN INVENTOR"

INVENTION SUBMISSION FORM

ATTACH PHOTO OR DRAWING OF INVENTION HERE (if available)

1. Name of Person Completing this Form: DANIEL B. HOCK
2. Are you a: ☒ Sole Inventor ☐ Team Member ☐ Other: ______ (please indicate)
3. Please list the name(s) of all inventor(s) who are submitting a Participant Application in connection with this invention, including yourself (attach the completed Participant Application(s) to this form):

DANIEL B. HOCK

4. Name/Title of Invention (if any): Monitored Felon Warning System
5. Description of the invention (include the intended end user and the purpose of the invention):

The present invention comprises a monitored felon warning system for allowing a user to get information of approaching monitored felons. The system comprises a receiver to use globally, without a license, the supplied frequency.

6. Date of conception of the invention: March 2005
7. What was the inspiration for the invention? During our retirement destination move, consideration was given to the safety of our children. More populated move, concerned for child molesters. Research has shown an increase. Devise a system to protect children. Increased concern for terrorism.

Patent Pending Listing to Investors

This patent description was originally listed on an auction site in January of '06, while patent pending, to request investors. Our patent was pended in October of 05, and made public accessible. This was the beginning of public patent sniffing corporations. The beginning of their government contracts began. An American business contact with a Corporation dragged on for many years without any outcome for our involvement. This American Corporation received government contracts. I believed many involving our patent.

Please read on:

Too many screams and family nightmares are a plague to our nations children and anybody concerned about their surrounding safety. I am a professional life rescuer, inundated with these nightmares and blood curdling screams of terror, to be rescued. This became a lifetime goal.

Our fully patented device can relieve governmental, public, military, and court-ordered cases by this affordable device. This immediate auto ID proximity alert from satellite monitoring is needed. This enhanced, GPS tracking of pedophiles, felons, stalkers, terrorists and persons in need of informational interest is of great importance.

When a cellular activity is added, an immediate picture and text, of persons of interest will be displayed. Auto ID, fully marketable and legally approved. This device is fully patented giving you an immediate automatic warning ID, with proximities of 350 feet, RFID, and further, cellular GPS. "Safe zones" with info. This is a public device, which is military, capable and is in much needed over the current monitoring. Constant downsizing gives offenders more freedom. Authorities, military, and the public's children need this now! How many more lives will be lost?

Our schools, libraries, malls, theaters, etc. and not to mention the stalking of our parks being done by these felons, whom cannot be rehabilitated. Adding safety to military personnel is now achievable. Add auto tracking to all auto ID, microprocessor cellular devices, into the hands of all terrorists, by courier delivery, and now they are GPS traceable.

This patent covers all tactical alerts. Also, notification to individuals will give a "SAFE ZONE" for our "GRIPPED in FEAR" nation. Put the automatic alerts into the public awareness, IMMEDIATELY. Not AFTER THE FACT. A great reward would be to see probable victims saved.

IMMEDIATE authority, parent notification, and master computer, receives and relays pertinent information, IMMEDIATELY!! On every day, every city, another abduction.

Over 1 million satellite dedicated, monitored pedophiles are under supervised, also, along with our terrorists (Until our patent sniffed).

NO MORE ALIBIS WITH THIS SYSTEM

A very lucrative, public affordable safety device estimated at $25 and up, with enhanced fees: data and text. The base model is a key fob RFID receiver, which reads a felon ankle transmitter.

With our patent, allowances for military, border protection, and terrorism gives IMMEDIATE troop notification through GPS information satellite phones. Thus, giving much-needed information through a parent master computer. GPS auto ID, cellular handhelds, many to be carried by every service personnel, Coast Guard and all authorities. We can now stand down terrorism with border protection and abroad.

Our current monitoring companies are not an immediate warning. Our probation officers need help. This device will help! Our children's lives are at an IMMEDIATE SAFETY ALERT! EVERY SECOUND COUNTS! Pennies a day can save so many lives. Military and border protection will progress. The countless abduction victims should be remembered. Their deaths have not and continue to grow DAILY. GOD BLESSES. PLEASE HELP.

More murderers, homeless, tier 3 predators, roaming freely, as addressed, IN THE AREA. This causes more needless fear of untraceable, monitored felons. ALERT THE PUBLIC. A SAFE ZONE is DETRIMENTAL.

Our device is working today. From inception, starting patent pending, of October 2005. Fully patented in November 2007. Many military conflicts overseas will give our patent great importance.

In this book you will read of many timelines and propositions offered but never received. No other patents, to this date, are known or approved in the United States to offer this personal, public or military protection, until now.

The Stress Memory

This timeframe started to become very stressful. We relied on much enjoyment as beaching and boating could. Living close to the coast was fantastic. Our stress was becoming stronger. My disability and pain was laying strong on our family. We were still in contact with more companies to join in our passion to protect.

Our hardships, beginning in 2003, were ending, and 2004 was taking off as a booming year. Finding we jumped too quickly at our chance to have a wonderful new home near the beach, it did not take long to find our dream home was going to be this hard. Continuing negative feedback and losing disability compensation was a tremendous stress. I can say 2006 was the end, to a belief, of a new beginning. We began to look hard at our finances, and the outlook of the economy, was in shambles. Great. What the hell happened!

Looking back between 2004 and 2006, our lives were flying at 1000 miles an hour. With the loss of my compensation and the negative outlook of our patent, there was a blessing in disguise. If we had not packed up and sold our new home by the beach, we would have been in bankruptcy. Our stars lined up, we sold quickly, and did not start the foreclosure meltdown.

It looked like the Hock family was flying the coop. In the beginning of 2007 we returned home to Ohio. Wonderfully, I was discussing with Carol, of another avenue to have our patent noticed. The Monitored Felon Warning System is our patent name. If it were not for this move, back to Ohio, we would have been in crappy, bad shape.

If we were to promote our patented device, my next step would be to incorporate some type of business. OFFENDAWAY, LLC. With the help of my accountant, my wife and I, were a Corporation. Now what? I needed more in-depth thinking. Carol would always say that I was just watching the grass grow. I love her so much!

Nightmares from my past career had driven me harder to succeed. Out of one, sky-blue day, I contacted WRGT-TV Fox 45. I discussed our patent with the editor after watching a news clip about pedophiles. We were amazed of the outcome. A news crew showed up at our door the next day and wanted to do a news clip about our invention. We saw ourselves on television that evening. Malcomb and Fish, did a wonderful, proud of, news clip. I cherish and share this clip. It wasn't 15 min. of fame, but at least our family was involved.

We were mind driven of this involvement. This is what we were destined to achieve, FIRST and SHARE, not to be, American

Corporation deceived. We now may be recognized. Can you believe that in 2 days the American Corporation contacted us?

AN UGLINESS MADE Its APPEARANCE

By naming this American Corporation my teeth grind together. I can tell you this American Corporation was in control of our destiny. This American Corporation had us very excited and quickly invited us for a business meeting. Our news clip was only aired 2 days ago! A main contact man was so overly nice that my insight was telling me something was fishy. This happened on July 29, 2009. I could say this was my day of infamy. Carol and I were still delighted that we had some attention. Finally, maybe, our protection device will be, at least, locally known.

I was a little concerned at the meeting and noticed a man so smugly leaning back in his chair with his long hair and beard. His demeanor seemed to me, as he was a king. Our recordings began, our non-disclosures were signed, and our contacts with American Corporation began.

I had no idea of the deceiving and deception that was to follow. There were misleading's in our recordings. We were directed away from military involvement. I was mind struck to their mis-wording. I believe there was a lot of misrepresenting of American Corporation happening. I believe American Corporation had done their job of deceiving. This was my beginning of HATE.

Contact of our Main Man

This American Corporation gave me a contact. This person I considered a Main Man. It made me feel so comfortable that I had such a trusting person that I could relay a lot of feelings to. My wife was not so trusting. I believe this is why we belong together. I am so naïve and Carol, well; I love her as my soul mate. For me, to believe, as such a person could exist, is probably why I got STOMPED ON. This is not what I believed a main man would do. I could've told him anything and believe that I was being heard. I trusted Main Man. You will find out that this man was no man to be trusted. Carol gave me full contact with him. The continuing years will become a huge never-ending roller coaster ride.

Still, without investigating American Corporation, daily contacts were made. This was the beginning of the wild roller coaster ride. A previous contact, while residing in Florida, was contacted again using American Corporation credentials. Numerous e-mails and promises of business contracts lead us into a 2nd NDA. I was so positive that we were on our way to fulfilling our "destiny to protect". Contacts would continue in and out over the next 2 weeks. I lost my 1st contact friendship. I believe my continued interest was to distract me from the construction of our patented unit for military use and contracts.

"This was just for the East Coast". "We still had contact for the West Coast". I believe more stalling. I now have a degree in DECEPTION. "Dan we are looking at something big". "Dan this could be what you wanted". "Let's make contact with West Coast". The 2nd contact was also a previous friendly interest.

One company controls the East Coast parole violator visits. The 2nd company controls the Western sectors. I had been in peaceful discussions with both East and West.

E-mails were back and forth between Main Man and I. This contact with the West Coast went very heated. I would receive copies of Main Man e-mails with West Coast. Eventually, over weeks of heated discussions with West Coast, I was at another loss. Disturbingly, I was very reluctant to discuss any more letdowns with my wife. She was fed up. Main Man was determined to keep my interest or wreck my life. I believe the latter. He would always come up with new avenues of possibility to excite me. There was nothing positive. "Let me contact my China man." All right. Can you believe I still believed? Was this more stalling? The disadvantaged American Corporation was digging in its heels.

Within these time frames, all e-mails and contacts that were made by Main Man, I have compiled quite a few, a lot. Finally, my clouded brain reverted to some of my avenues. My main Avenue was to investigate American Corporation. From out of the clouds, I learned quite a bit from their 1st website. I began contact for help. I had to understand if my misleading was grounds for infringement. It really didn't matter because there were so many other American companies doing the same thing. Auto GPS tracking was taking off. Auto ID, locator by a cellular information display was built and being promoted to the military and government.

I would like to go back to our 1st meeting with American Corporation and remember of the gentleman with long hair and beard. He had just landed a $500 million contract. Let's see, we were patent pending in October 2005, and fully patented in November 2007. These public records of patents can be read at patent pending levels. It has taken me too long to put 2+2 together.

Our company, OffendAway, LLC, had been locally televised on a local news Channel. After viewing this news clip, with my belief of the timelines, and investigating American Corporation and their government contracts was overwhelming.

By this time, over $900 million had been awarded to this disadvantaged American Corporation where I have related my complete trust. To this day I do not understand why I was so left out. This awarded contract has expeditiously propelled this American Corporation. As my STRONG belief, this patented microprocessor device had been promoted to the government and military. This extremely large contract was awarded within weeks of the airing of our news clip.

(Pat.pend-'05, sniffed then contracted in '09)

Many millionaire individuals have been lifted to riches nearly overnight. The deception of this American company and their contact with my previous interests, has denied me any further advancement with my friends. Today, my family is still disadvantaged, my being a disabled firefighter, "bread and butter", and dependent on disappearing finances. I'm not Laughing Out Loud. More like S.O.L., is this our democratic government, to call, "your on your own, winner take all"?

My American Corporation obviously had read our patent to protect and put 2+2 together. They had found our patent for auction, cellular GPS, auto ID, auto tracker, and proposed it to the military. This is my strong belief. My insight does not fail me.

Classified: Bin Laden Goes Down

I had learned not only $500 million in contracts but nearly 1.5 Billion dollars in contracts were awarded to American Corporation. The other large corporation involved also had an interest in this endeavor; I can say something to do with my cell phone. I was wondering about this American Corporation on my cell phone and sent them a letter. I was curious if they had any involvement with my patent. Of course they denied it. I did not have proof. This was more stalling for American Corporation.

Paperwork was flying in our home. My stress level was elevated. THE PAIN, PAIN, NO RELIEF. To just start investigation of a possible infringement case was $100,000. No freebie from New York. I still have faith, and hate.

Dan Hock
9498 Ash Hollow Ln.
Centerville, Oh.45458

[redacted] [redacted]y,Inc.
Law Department
[redacted] North U.S. Highway [redacted]
Libertyville, IL [redacted]

RE: Hock, Dan Inquiry To [redacted] Ethics Line on July 29,2010
Response:

[redacted]
Lead Intellectual Property Counsel,

It has come to my attention, after my expired Non-Disclosure , that a contract valued in millions of dollars has been awarded. Due to your handheld [redacted] 9090 , there has become an interest in. With my patent (# 7289031) and your tracking, monitoring and information database sharing , I would like your review , ask for schematics and processes of your unit.

My patent was awarded with claims of database monitoring, tracking and information sharing including felons, others of interest , whether military and public awareness. Without an Attorney at this time, I am sure we can discuss these claims, ethically , and in my view avoid possibly litigation of infringement.

With your strong view of respect for intellectual property, I am sure we will reach licensing opportunities.

Keeping it simple and with very sincere thoughts,
Dan Hock
(937) 232-3933
Dhock@who.rr.com

August 31, 2010

VIA US MAIL AND E-MAIL
www.Dhock@who.rr.com

Mr. Dan Hock
9498 Ash Hollow Lane
Centerville, OH 45458

Re: US Patent No. 7,289,031

Dear Mr. Hock:

This is in response to your letter of August 24, 2010. I will be the attorney handling this matter in-house. Please direct all correspondence regarding this matter to my attention.

In [redacted] letter of August 11, 2010, [redacted] requested identification of the patent that is the basis of your infringement allegations and information on how you believe this patent is relevant to [redacted] business. While you identified U.S. Patent No. 7,289,031 in your August 24 letter, you have not provided [redacted] with the basis of your infringement allegation.

Based on our initial review of the '031 patent, we do not understand the basis of your allegation that [redacted] 909[redacted] infringes the claims of this patent. Thus, we reiterate our request that you provide [redacted] with a specific explanation that is the basis of your assertions involving the applicability of the claims of the '031 patent to [redacted] 9090.

Sincerely,

[redacted]

With my continuing investigating timelines involved, following e-mails, and American Corporation, my insight was working overtime. I know now, that back then, July 2009, was the beginning to the end of our American Corporation involvement with our patent, The Monitored Felon Warning System.

Within our patent, it is described, all passive devices to alert covers cell phone ID and location warnings. Cell phones connect by satellites, and signals are bounced around the world. Our GPS auto tracker auto ID microprocessor could now be implemented into cell phones. Anybody can be tracked, located and IMMEDIATELY identified. 4G?

In 2009, everybody in America knew who our worst enemy on the planet was, Osama bin Laden. This man has eluded the Soviets for over 10 years. This man has devastated our country with his attacks and has continued to elude even 2 US presidencies.

I was following the news closely, and seemingly contacts with Main Man had slowed down. In 2009, I had a surgery on my back to fuse discs together. It was a very hard time. My mind was not with it. Main Man knew of this time because I still trusted him.

While 2010 rolled around with economy worsening, finances, everybody was on edge. I know it and you know it. There were no royalties from our invention or business contracts. All promises were never fulfilled. The only thing I know, is that American

Corporation has grown tremendously and leaving town. I had promised that I would not leave town with that agreement.

With our system, an American corporation added a new name for it, auto ID. While, I can take a Coke and rename it. It is mine now. Where are my contracts? While I was recovering, our patent was being taken advantage of. I went back to American Corporation website and noticed that it was changed. All persons in our initial business meeting had looked as though they had total makeovers. These makeovers must have cost tens of thousands of dollars. They all seem to have gained presidential status.

Big news has happened while I was recovering. In my years I have had a total of 7 knee surgeries, 4 back surgeries and constant pain that had distracted me. I do know that many terrorists, murderous dictators, were dropping like flies especially our man Osama Bin Laden.

I was elated and disturbed altogether. This was definitely a quick and quiet operation. I went totally rewind. There were stalling e-mails. There was no need for an international patent. I do believe a company involved is and OVERSEAS INTERNATIONAL CORPORATION. Why can't I get this out of my head? I guess, one man with an insightful, patented, innovative device, of this locating magnitude, could not have assisted in this operation at all.

I did e-mail the White House. I had a wonderful return. E-mail thanks for your suggestion. Please continue to invent. A presidential e-mail of thanks not even signed. Thanks Mr. President. Our disadvantaged American Corporation contracts should also include, totally disabled small business owners, trying to make a mark. This wayward deception must stop. I should have some notoriety to my name. I guess I will frame my presidential e-mail.

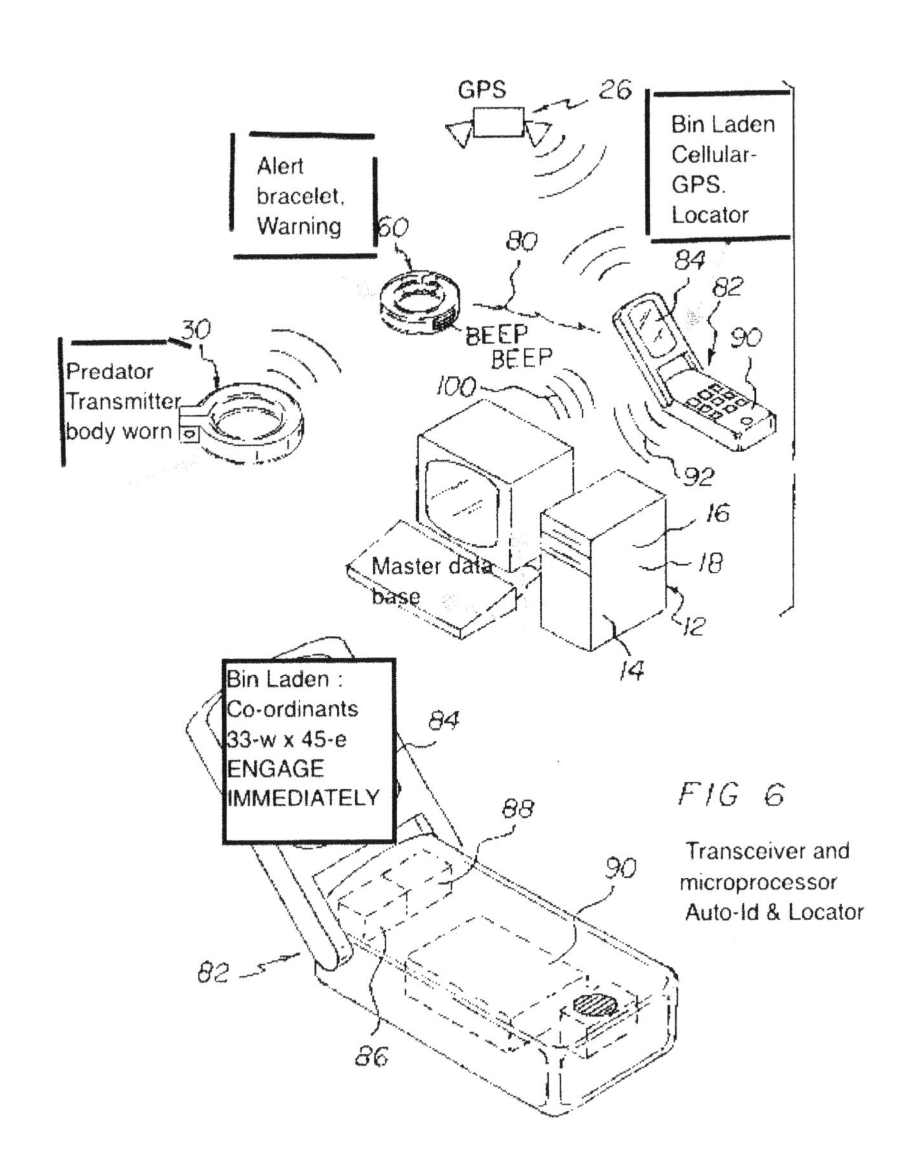

FIG 6

Transceiver and
microprocessor
Auto-Id & Locator

RoadRunner Search

Read Message

Usage 70% of 1000.0 MB

Compose Get Mail Search Mail Address Book Settings Log Out

Inbox (8)
Drafts (8)
Sent Mail
Deleted Items (146)
Junk Mail

Reply Reply All Forward View Header Delete Report SPAM Printable View Move to:

<< Prev | Next >>

dhock001@woh.rr.com

ch.com>

Subject: Thank YOU

Priority: Normal **Date:** Thursday, October 15, 2009 11:46 AM **Size:** 3 KB

Manage Folders
apple
bestbuy
infringe
paid
paid off
past
paypal
printed

----- wrote:
> Dan,
>
> I don't think you need to worry about an international patent at this time. Perhaps as we move through prototyping it would be value. My personal take is that most countries are willing to have ideas taken and they do little to enforce anyway. That said, I leave the call to you. — Mislead
>
> Our technology team had its first meeting, and I have briefed our and . We will be hosting another call with shortly to explore technical questions. I would also expect that our team will create a teaming agreement for a joint venture to market the application.
>
> Have a a great weekend and I shall keep in touch,
>
>
>
>
>
> -----Original Message-----
> From: dhock001@woh.rr.com [dhock001@woh.rr.com]
> Sent: Friday, October 09, 2009 10:20 AM
> To: ,
> Subject: International patent?
>
> ,
> I should have your # again when my son returns my sim card. Should I be concerned about an international patent?
> I am deeply indebted to your desire and help to bring safety to our Nations' children. Words cannot describe my feelings...THANK YOU and (GOOD THOUGHTS always)
> Daniel B. Hock
> Offendaway,LLC
> ----- wrote:
> > ,
> >
> >
> >
> > Last Friday we had an introductory meeting with Dan Hock (no company)
> > regarding a patented concept he holds and the potential for to
> > assist in bringing the concept to market. So that we may explore a
> > relationship further, we agreed to enter into a mutual NDA. Could you
> > please forward such to Dan (copied on this note) and then let me know
> > when it is in place?
> >
> >
> >
> > Thanks so much,
> >
> >
> >
> >
> >
> >
> >
> >
> >
> > Director, Commercial Marketing
> > Technologies, INC.
> >

1 of 2 9/2/12 4:57 PM

Really, a good thing did turn out by moving back to Ohio. My disability compensation was returned and some happiness came with it. My wife loves me. My children love me. My attention and depression is still lagging. I wish I could tell them how much they truly mean to me. My wife has stuck by my side through my worst times and I know I still have more to come. The pain in my back and aggravation of being robbed disrupts my thoughts daily. My dignity, my self-worth, has diminished. In my heart I know what we have accomplished.

Hussein-2006, Gaddafi-2011, Bin Laden-2011
Officials keep knocking on my door.

A Most Influential Person (not me)

Believevably, American corporations have specialists only hired to search patents. New patents. Our new patent. This American Corporation owner is now coveted awarded as a most influential person. Since American Corporation learned of our patent, they started on their own endeavor, creating overnight success. With a $1.7 million award to build 780 auto identification integrated cellular microprocessor locators. This was awarded April 6, 2010. Many other American corporations were included. Weeks after weeks, during the year 2010, brought American Corporation billions of dollars in government and military contracts.

(Offendawayllc inventor, NOTHING BUT HEADACHES)

In May of 2009 there was a $428 million contract awarded. This was only for the military and Coast Guard. In March of 2010, April 2010, more contracts. It seemed as though we were missing many contracts and royalties. Imagine, contracts over $1.5 billion especially, after the death of bin Laden, Hussein, Gadhafi, and continuing, with high-ranking Al Qaeda terrorists located daily. Was this due to GPS tracking drones? I just can't believe, a man, suddenly located and dispatched, after decades of multiple country searches, absolutely, nearly overnight, out of the blue, located.

In March of 2005, a tracking bill was introduced into legislation that would require all felons to wear a tracking device. This would be in the form of RFID and with the increased technology of cellular auto ID, auto tracking, GPS monitoring, our wonderful device has been improving. Obviously, this was being done under my nose and behind my back. This was the beginning of our direction to continue to protect.

Now that the military and Coast Guard has this cellular integration, again, our involvement, has the increased ability, destined to improve implementing GPS, auto ID tracking microprocessors into the public mainframe. Finally, our smart phones will have the auto ID tracking systems to help avoid danger. WE ARE PROUD. PROTECT OUR CHILDREN.

Now that this miracle device has redeemed our faith in our military protection, we can now, hopefully, protect our country. Protect our people of the GREATEST NATION ON EARTH! Protect our borders. Why stop there? Why not stop the ABDUCTION, TORTURE of the heinous MURDERS of OUR CHILDREN? It can be achieved. If our American corporations would stop thinking about when my patent will expire maybe this protection will start. My mind is set on completing our goal.

(12) **United States Patent**
Hock

(10) **Patent No.:** **US 7,289,031 B1**
(45) **Date of Patent:** **Oct. 30, 2007**

(54) **MONITORED FELON WARNING SYSTEM**

(75) Inventor: **Daniel B. Hock**, Seminole, FL (US)

(73) Assignee: **Carol G. Hock**, Centerville, OH (US)

(*) Notice: Subject to any disclaimer, the term of this patent is extended or adjusted under 35 U.S.C. 154(b) by 204 days.

(21) Appl. No.: 11/290,604

(22) Filed: Nov. 30, 2005

(51) **Int. Cl.**
G08B 23/00 (2006.01)

(52) **U.S. Cl.** 340/573.4; 340/573.1; 340/572.1; 340/539.13; 455/456.1

(58) **Field of Classification Search** 340/573.4
See application file for complete search history.

(56) **References Cited**

U.S. PATENT DOCUMENTS

5,396,227 A * 3/1995 Carroll et al. 340/825.36
5,982,281 A * 11/1999 Layson, Jr. 340/573.4
6,014,080 A * 1/2000 Layson, Jr. 340/573.4
6,054,928 A * 4/2000 Lemelson et al. 340/573.4
6,072,396 A * 6/2000 Gaukel 340/573.4
6,100,806 A * 8/2000 Gaukel 340/573.4
6,362,778 B2 * 3/2002 Neher 342/357.07
6,639,516 B1 * 10/2003 Copley 340/573.4
6,674,368 B2 * 1/2004 Hawkins et al. 340/573.4
6,828,908 B2 * 12/2004 Clark 340/573.1
6,889,135 B2 * 5/2005 Curatolo et al. 340/572.1
6,972,684 B2 * 12/2005 Copley 340/573.4
7,015,817 B2 * 3/2006 Copley et al. 340/573.4
7,098,795 B2 * 8/2006 Adamczyk et al. 340/573.4
7,102,508 B2 * 9/2006 Edelstein et al. 340/539.13
7,123,141 B2 * 10/2006 Contestabile 340/573.4
2002/0063626 A1 * 5/2002 Pitzer et al. 340/573.1
2005/0099309 A1 * 5/2005 Hum et al. 340/573.4
2005/0285747 A1 * 12/2005 Kozlay 340/573.4
2006/0063540 A1 * 3/2006 Beuck 340/539.13
2007/0069890 A1 * 3/2007 Tuck 340/539.13

* cited by examiner

Primary Examiner—Benjamin C. Lee
Assistant Examiner—Eric M. Blount
(74) *Attorney, Agent, or Firm*—Edward P. Dutkiewicz

(57) **ABSTRACT**

A monitored felon warning system has a police computer having a felon data base. The system also has a global positioning system, also known as GPS, and a sending subassembly having a circuit, with the transmitter being capable of transmitting a first signal to a receiver. The receiver gives off an alarm and also causes a display of warning material, such as a photograph to be displayed on the cell phone screen.

7 Claims, 5 Drawing Sheets

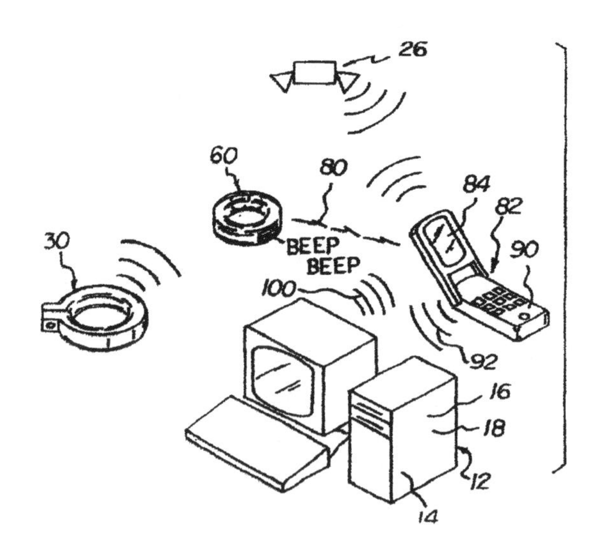

No Coincidence

Our fully patented and proven device has given detrimental innovations to the government. By deceptive American corporations of our groundbreaking technology, depriving our small business from claiming any involvement of our "destiny to protect". These insightful rights to innovate, invent, patent, and produce only to be deceived by large American corporations with promises of success. We then were eliminated. Given, without concern, our device to an American corporation of United States Air Force Defense Department.

Military and Coast Guard contracts were awarded to them. Our initial televised news clip of our device on July 26, 2009, and as of October 2005 we were patent pending. From the attack of Al Qaeda in 2001, terrorism attacks were daily major headlines. Al Qaeda and numerous abductions and horrendous torture of our American troops were televised and placing the United States into a constant feeling of despair. Untracked Al Qaeda terrorism was a major concern. Our patent, I am most assured, returned a new direction to ANSWER THE CALL of AMERICAN RECOURSE.

Decades searching for these terrorists were unsuccessful. This hunt for Al Qaeda continued, until now.

Dispatching terrorism, on every corner of the globe, is the answer. Our device, cellular GPS auto tracking auto ID felon monitoring microprocessor does the positive recourse to action. I have dreamed for decades of my worth. I have dedicated myself for decades to reach this goal. Our dream of innovation and our insight is to be fully patented. This is no coincidence. Terrorism is being controlled. Our Troops of Success have patriotically returned home. Is this my worth?

If this American Corporation did not mislead me and my nondisclosure agreements were followed, my dreams would've been true. I still cannot shake the hate. I don't believe I ever will. Our patent may expire in 10 years, but my dedication to protect, and my self-belief, will never give out.

By the year 2015, our government plans to put into our skies thousands of drone planes. Unmanned to patrol our skies. Our microprocessor? This cellular auto locator will be locating our public? At a military standpoint, the unmanned drones can be armed with missiles, and possibly warheads that could be released onto any country that deems a threat, which is very good. I hope this only happens overseas. In the public standpoint, personally, I would wish to know if my cell phone is or is not being tracked. My guess is, it is.

This device has come a long way in a very short time. Our invention was not derived in a garage, but in the concerns of my wife and I, having a dedication to protect. Drones may soon be in a sky near you.

Maybe our next step would be a discussion to personally microchip and barcode to assure citizenship. Thank you.

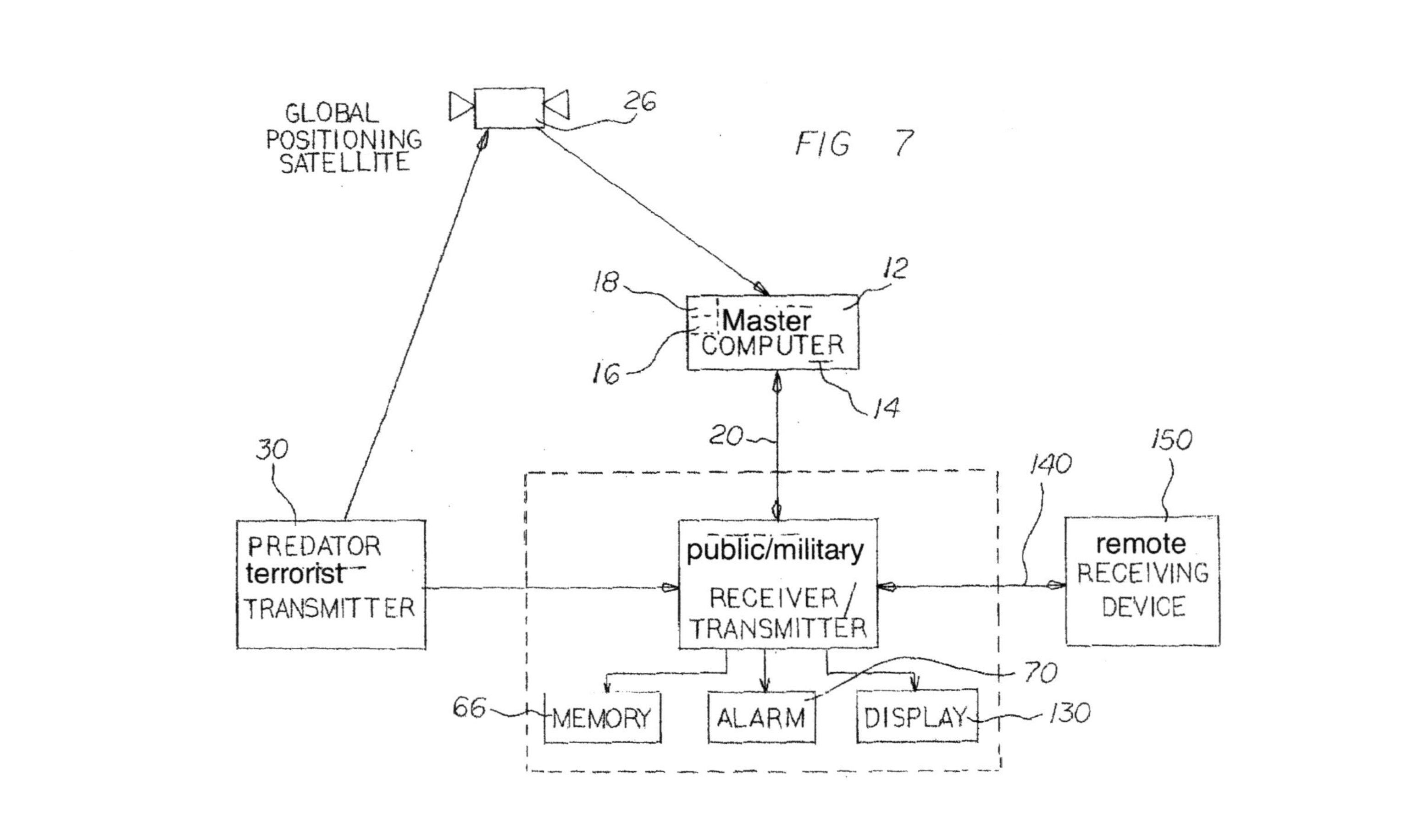
GLOBAL POSITIONING SATELLITE
26
FIG 7
Master COMPUTER
12
18
16
14
20
30
PREDATOR terrorist TRANSMITTER
public/military
RECEIVER/ TRANSMITTER
140
150
remote RECEIVING DEVICE
70
66
MEMORY
ALARM
DISPLAY
130

Low frequency and high-frequency RFID tags can be used globally without a license. Ultrahigh frequency cannot be used globally as there is no single global standard. In North America, UHF can be used unlicensed for 908-928 MHz, but restrictions exist for transmission power.

ISO 15693 is an **ISO** standard for "**Vicinity Cards**", i.e. cards that can be read from a greater distance as compared to **Proximity cards**.

ISO 15693 systems operate on 13.56 MHz frequencies, and offer maximum read distance of one-1.5 m. An example of this being the radio identification tags, RFID, used to collect toll electronically these days.

As the vicinity cards have to operate at a greater distance, the necessary magnetic field is less than that for a proximity card.

ISO 14443 defines a proximity card used for identification that usually uses the standard credit card form factor defined by ISO 7810 ID-1. Other form factors also are possible. The ID cards use an embedded microcontroller, including its own microprocessor and several types of memory, and a magnetic loop antenna that operates at 13.56 MHz (RFID). More recent ICAO standards for the machine-readable travel documents specify a cryptographically signed file format and authentication protocol for storing biometric features such as photos of the face, fingerprints, and/or Iris.